READING ABOUT SCIENCE
Skills and Concepts

BOOK A

John Mongillo
Beth Atwood
Kevin M. Carr
Linda J. Carr
Claudia Cornett
Jackie Harris
Josepha Sherman
Vivian Zwaik

Phoenix Learning Resources, LLC

PHOTO CREDITS

COVER PHOTO—Steve Vidler/SuperStock; **2**—Geroge Porter/Photo Researchers;
5,10—Firefly Productions/The Stock Market; **5,26**—National Zoological Park; **5,16**—Alan Briere/Superstock;
5,28—Anita Sabarese; **6**—Jen and Des Bartlett/Photo Researchers; **8**—San Diego Zoological Society;
12—Tom McHugh/Photo Researchers; **14**—Hugh Spencer/Photo Researchers;
18—Jen and Des Bartlett/Photo Researchers;
20—Jack Donnelly/Woods Hole Oceanographic Institution, Dr Kathleen Crane/Woods Hole Oceanographic Institution;
22—U.S. Department of Agiculture Center for Nutrition Policy and Production;
24—Christa Armstrong/Photo Researchers; **30**—Burpee Seeds; **32**—Anita Sabarese;
34—NASA; **37,42**—Superstock; **37,50**—American Association for the Advancement of Science;
37,38—H.Armstrong Roberts; **37,54**—Gerard Fritz/Superstock; **40**—Alan Pitcairn/Grant Heilman;
44—Superstock; **46**—United States Air Force; **48**—Dennis Scott/The Stock Market;
52—John Porteous/Woods Hole Oceanographic Institution; **56**—Mimi Forsyth/Monkmeyer;
58,60—Joan Lifton/Woodfin Camp & Associates; **59,68**—Robert Capece/McGraw-Hill;
59,74—Robert Capece-McGrae-Hill; **59,62**—Betty Adams/Monkmeyer; **59,70**—Keith Ross/Superstock;
64—Superstock; **72**—Thomas Raupach/Woodfin Camp & Associates; **76**—SuperStock;
78,90—Mary Ann Kulla/The Stock Market; **79,88**—Superstock; **79,86**—H.Armstrong Roberts;
79,92—Chris Sorensen/The Stock Market; **79,94**—Peter Beck/The Stock Market; **80**—H.Armstong Roberts;
82—Steve Vidler/Superstock; **84**—Derek Trask/The Stock Market; **96**—Kennan Ward/The Stock Market.

Phoenix Learning Resources, LLC

914 Church Street • Honesdale, PA 18431
1-800-228-9345 • Fax: 570-253-3227 • www.phoenixlr.com

Item# 2201   ISBN 978-0-7915-2201-1

Copyright © 2010, 2001, 1990, 1981 Phoenix Learning Resources, LLC., All Rights Reserved.
This book is not to be reproduced in any manner whatsoever, in part or whole,
without the written permission of Phoenix Learning Resources, LLC.

Authors

John Mongillo, Senior Author and General Editor
Science Writer and Editor
Saunderstown, Rhode Island

Beth S. Atwood
Writer and Reading Consultant
Durham, Connecticut

Kevin M. Carr
Teacher and Writer
Honolulu, Hawaii

Linda J. Carr
Writer and Psychologist
Honolulu, Hawaii

Claudia Cornett
Professor Emerita
Wittenberg University

Jackie Harris
Medical and Science Editor
Wethersfield, Connecticut

Josepha Sherman
Writer and Science Editor
Riverdale, New York

Vivian Zwaik
Writer and Educational Consultant
Wayne, New Jersey

CONTENTS

Page

Life Science 2
    A Living Dragon 6
    When the Bears Turned Green 8
    How Smart Are Animals? 10
    The Lungfish 12
    Hide and Seek 14
    Orcas 16
    Ants That Look for Leaves 18
    A Strange World Under the Sea 20
    Think Before You Eat 22
    Your Teeth and You 24
    Black and White All Over 26
    Meet the Cactus 28
    Sugar Snap Peas 30
    The Ginkgo Tree 32

Earth-Space Science 34
    Spinning Winds 38
    A Black Graveyard for Animals 40
    Moon Secrets 42
    A Nosy Neighbor 44
    Hurricane Watch 46
    Meteorites 48
    Mountains of Ice 50
    The *Alvin* 52
    What Did Dinosaurs Do? 54

	Page
Physical Science	**56**
Hang Gliding	60
Hot-Air Balloon	62
Lasers in Medicine	64
What Do You Think?	66
How a Battery Works	68
Dangerous Rays	70
Maglev: The Flying Train	72
The Uses of Petroleum	74
Environmental Science	**76**
The Bald Eagle Is Soaring	80
Solar Heat	82
Fuel for Tomorrow	84
Acid Rain	86
We Need a Pollution Solution	88
Helping Mother Nature	90
Charged-up Cars	92
Bottle Bills Reduce Waste	94
Mountain Gorillas and Dian Fossey	96
Bibliography	**98**
Record Keeping	**100**
Metric Tables	**103**

TO THE STUDENT

Do you enjoy the world around you? Do you ever wonder why clouds have so many different shapes and what keeps planes up in the air? Did you ever want to explore a cave or find out why volcanoes erupt or why the earth shakes? If you can answer yes to any of these questions, then you will enjoy reading about science.

The world of science is a world of observing, exploring, predicting, reading, experimenting, testing, and recording. It is a world of trying and failing and trying again until, at last, you succeed. In the world of science, there is always some exciting discovery to be made and something new to explore.

Four Areas of Science

READING ABOUT SCIENCE explores four areas of science: life science, earth-space science, physical science, and environmental science. Each book in this series contains a unit on each of the four areas.

Life science is the study of living things. Life scientists explore the world of plants, animals, and humans. Their goal is to find out how living things depend upon each other for survival and to observe how they live and interact in their environments, or surroundings.

Life science includes many specialized areas, such as botany, zoology, and ecology. *Botanists* work mainly with plants. *Zoologists* work mostly with animals. *Ecologists* are scientists who study the effects of air pollution, water pollution, and noise pollution on living things.

Earth-space science is the study of our Earth and other bodies in the solar system. Some earth-space scientists are *meteorologists*, who study climate and weather; *geologists*, who study the earth, the way it was formed, and its makeup, rocks and fossils, earthquakes, and volcanoes; *oceanographers*, who study currents, waves, and life in the oceans of the world; and *astronomers*, who study the solar system, including the sun and other stars, moons, and planets.

Physical science is the study of matter and energy. *Physicists* are physical scientists who explore topics such as matter, atoms, and nuclear energy. Other physical scientists study sound, magnetism, heat, light, electricity, water, and air. *Chemists* develop the substances used in medicine, clothing, food, and many other things.

Environmental science is the study of the forces and conditions that surround and influence all living and nonliving things. Environmental science involves all of the other sciences-life, earth-space, and physical.

If you want to know more about one or more of these areas of science, check the bibliography at the back of this book for suggested additional readings.

Steps to Follow

The suggestions that follow will help you use this book:

A. Study the photo or drawing that goes with the story. Read the title and the sentences that are printed in the sidebar next to each story. These are all clues to what the story is about.

B. Study the words for the story in the list of Words to Know at the beginning of each unit. You will find it easier to read the story if you understand the meanings of these words. Many times, you will find the meaning of the word right in the story.

When reading the story, look for clues to important words or ideas. Sometimes words or phrases are underlined. Pay special attention to these clues.

C. Read the story carefully. Think about what you are reading. Are any of the ideas in the story things that you have heard or read about before?

D. As you read, ask yourself questions. For example, "Why did the electricity go off?" "What caused the bears to turn green?" Many times, your questions are answered later in the story. Questioning helps you to understand what the author is saying. Asking questions also gets you ready for what comes next in the story.

E. Pay special attention to diagrams, charts, and other visual aids. They will often help you to understand the story better.

F. After you read the story slowly and carefully, you are ready to answer the questions on the questions page. If the book you have is part of a classroom set, you should write your answers in a special notebook or on paper that you can keep in a folder. Do not write in this book without your teacher's permission.

Put your name, the title of the story, and its page number on a sheet of paper. Read each question carefully. Record the question number and your answer on your answer paper.

The questions in this book check for the following kinds of comprehension, or understanding:

1. *Science vocabulary comprehension.* This kind of question asks you to remember the meaning of a word or term used in the story.

2. *Literal comprehension.* This kind of question asks you to remember certain facts that are given in the story. For example, the story might state that a snake was over 5 feet long. A literal question would ask you: "How long was the snake?"

3. *Interpretive comprehension.* This kind of question asks you to think about the story. To answer the question, you must decide what the author means, not what is said, or stated, in the story. For example, you may be asked what the main idea of the story is, what happened first, or what caused something to happen in the story.

4. *Applied comprehension.* This kind of question asks you to use what you have read to (1) solve a new problem, (2) interpret a chart or graph, or (3) put a certain topic under its correct heading, or category.

You should read each question carefully. You may go back to the story to help you find the answer. The questions are meant to help you learn how to read more carefully.

G. When you complete the questions page, turn it in to your teacher. Or, with your teacher's permission, check your answers against the answer key in the Teacher's Guide. If you made a mistake, find out what you did wrong. Practice answering that kind of question, and you will do better the next time.

H. Turn to the directions that tell you how to keep your Progress Charts. If you are not supposed to write in this book, you may make a copy of each chart to keep in your READING ABOUT SCIENCE folder or notebook. You may be surprised to see how well you can read science.

PRONUNCIATION GUIDE

Some words in this book may be unfamiliar to you and difficult for you to pronounce. These words are printed in italics. Then they are spelled according to the way they are said, or pronounced. This phonetic spelling appears in parentheses next to the words. The pronunciation guide below will help you say the words.

ă	pat	î	dear, deer, fierce, mere	p	pop	zh	garage, pleasure; vision
ā	aid, fey, pay			r	roar		
â	air, care, wear	j	judge	s	miss, sauce, see	ə	about, silent pencil, lemon, circus
ä	father	k	cat, kick, pique	sh	dish, ship		
b	bib	l	lid, needle	t	tight		
ch	church	m	am, man, mum	th	path, thin	ər	butter
d	deed	n	no, sudden	*th*	bathe, this		
ĕ	pet, pleasure	ng	thing	ŭ	cut, rough		
ē	be, bee, easy, leisure	ŏ	horrible, pot	û	circle, firm, heard, term, turn, urge, word		
		ō	go, hoarse, row, toe				STRESS
f	fast, fife, off, phase, rough	ô	alter, caught, for, paw	v	cave, valve, vine		Primary stress ' bi·ol'o·gy
g	gag			w	with		[bī ŏl'ejē]
h	hat	oi	boy, noise, oil	y	yes		
hw	which	ou	cow, out	yōō	abuse, use		Secondary stress ' bi'o·log'i·cal
ĭ	pit	ŏŏ	took	z	rose, size, xylophone, zebra		[bī'elŏj'ĭkel]
ī	by, guy, pie	ōō	boot, fruit				

The key to pronunciation above is reprinted by permission from *The American Heritage School Dictionary* copyright © 1977, by Houghton Mifflin Company

LIFE SCIENCE

You have probably seen many frogs. But have you ever seen a giant tree frog? These amphibians are found in warm, humid climates. They grow to be about 5 inches long. (Some tree frogs are only 8 tenths of an inch long.) When the male frog sings, it makes a snarling, snorting sound. Imagine the noise when hundreds of tree frogs are singing!

WORDS TO KNOW

A Living Dragon
forked, having two tips
lizards, animals with a long slender body and tail, a scaly skin, and four legs, such as the gecko, iguana, chameleon, or salamander

When the Bears Turn Green
polar, near or from the North Pole
invisible, not able to be seen

Orcas
pod, a covering that holds several seeds; **also**, a group of seals or whales
slang, language of a particular group
salmon, a large fish with pinkish flesh
dolphin, a warm-blooded sea animal with a pointed nose

The Lungfish
shallow, not deep
curls, to bend around
necessary, must be done
straightens, unbends, uncurls

How Smart Are Animals?
instinct, behavior animals are born with
behavior, the way people and animals act, function
chimpanzees, apes with long black hair, long arms, and no tail. Very intelligent.
orangutans, apes with shaggy reddish-brown coat and no tail

A Strange World under the Sea
chemicals, the physical matter that makes up things
submarines, ships that can go under water (from *sub*, under, and *marine*, sea)

Think Before You Eat
nutrients, things that nourish, that feed life and growth

Your Teeth and You
dental floss, a thread to clean between teeth

Meet the Giant Cactus
extra, more
stem, the main above-ground stalk or trunk of a plant
usually, most of the time
type, kind of

Sugar Snap Peas
vegetables, plants in general that can be eaten or used for food
succeeded, to have success
tough, not easily chewed
raw, not cooked

The Ginkgo Tree
waxy, looks to be covered with wax
soot, black dust formed mainly by the incomplete burning of oil and gasoline

LIFE SCIENCE

Life Science

A Living Dragon

Are there really living dragons?

It is morning on the island of *Komodo* (kə mō′ dō) in the Indian Ocean. Out of a cave comes a giant animal. It sticks out a long forked tongue. Inside its red mouth are rows of teeth. Is this a dream? No! There is such an animal. It is called the Komodo dragon.

The Komodo dragon is a *monitor lizard* (mŏn′ĭ tər lĭz′ərd), the largest of all known lizards. It can grow as long as eleven feet and weigh as much as two people. The Komodo looks for food by smelling with its tongue. This giant lizard can eat pigs, deer, goats, and monkeys.

For hundreds of years, people have talked about dragons as make-believe. Then, in 1912, people heard about the Komodo dragon. They thought it too was make-believe. But first four Komodo skins and then photographs were brought back from the island. Then people knew that the Komodo dragon was real.

QUESTIONS

1. The *monitor lizard* is
    a. a small lizard with many rows of teeth.
    b. a make-believe lizard.
    c. the largest known lizard in the world.

2. The Komodo dragon uses its tongue to
    a. catch snakes.
    b. scare people.
    c. smell food.

3. People may not have believed the Komodo dragon was real because
    a. they had never seen one themselves.
    b. the dragon was not alive until 1912.
    c. the dragon never came out of its cave.

4. Which of the following happened *first*?
    a. People saw photographs of the Komodo.
    b. Komodo skins were brought back from the island.
    c. People knew that the Komodo was real.

5. According to the story, which of the following statements is *true*?
    a. All the stories about the Komodo dragon were make-believe.
    b. Sometimes what we think is make-believe turns out to be real.
    c. People who saw the Komodo dragon were dreaming.

When Bears Turned Green

Read all about the zoo bears that turned green.

Polar bears are supposed to be white. They live in the cold northern parts of the world. Their white fur makes them almost invisible in the snow and ice.

But three bears living in a zoo turned green. That's right! The big white bears turned as green as grass. Scientists wondered why it had happened. They cut away some of the bears' fur and looked at it carefully. There were tiny plants called algae (ăl'jē) inside the bears' fur. Algae are plants that grow mostly in water, and they do not have true leaves or flowers.

Some kinds of algae can be green, and it was this algae that changed the color of the bears' fur. Scientists believe the algae were living in the small pool used by the bears. When the bears went for a swim, the algae moved into the bears' fur. The algae were not hurting the bears. The bears were not hurting the algae. In fact, they lived together happily. But green polar bears were a strange sight!

QUESTIONS

1. *Algae* are
    a. a kind of bear.
    b. tiny plants.
    c. plants with flowers.

2. Algae live
    a. in the water.
    b. in dry places.
    c. only in zoos.

3. The bears turned green because
    a. they were sick.
    b. they went swimming too often.
    c. there were plants in their fur.

4. Polar bears can live safely in ice and snow because
    a. their white color makes it hard for hunters to see them.
    b. they can see hunters coming from far away.
    c. people do not hunt where there is ice and snow.

5. In this story, where were the algae before they got into the bears' fur?
    a. in the snow
    b. in the flowers
    c. in the pool

How Smart Are Animals?

What are some ways animals use tools?

Scientists once thought just people use tools. They thought animals were not smart enough. By watching them, scientists found out animals are very smart.

A tool is an object used for a task. Thinking is needed to use tools-not just instinct. *Instinct* is behavior animals are born with. They do not need to think about what they do by instinct.

Scientists have now seen many animals use tools. Chimpanzees use twigs to dig up bugs. First they find a twig. Then they strip off the leaves. Then they break the twig to the right length. Finally they stick the twig into termite holes. Termites that grab the twig are pulled out. Lunch!

Birds have also been seen using twigs to get insects. Orangutans, which are big apes, are good tool users, too. One orangutan bent a wire to open his cage. Then he escaped!

QUESTIONS

1. A tool is
    a. an object used to do a job or task.
    b. an object animals buy.
    c. an object made of metal.

2. When animals use tools, they are
    a. using instinct.
    b. thinking.
    c. not being very smart.

3. What is another animal that uses sticks to get insects?
    a. orangutan
    b. whales
    c. bird

4. What was the last thing the chimps did with their twig tools?
    a. broke them into the right length
    b. ate them along with the bugs
    c. stuck them in the termite hole

5. Which of these groups could best be used as tools by animals?
    a. bike, car, plane
    b. stone, stick, string
    c. ball, light, clock

LIFE SCIENCE

The Lungfish

Did you ever hear of a fish that can live for a long time without being in water?

The lungfish breathes through gills just as other fish do. But unlike other fish, it also has lungs. So the lungfish can breathe air just as animals living on land do. For this reason, a lungfish can live for a long time without being in the water.

Most lungfish live in Africa, where they are found in shallow lakes and rivers. Sometimes very little rain falls. Then the shallow lakes and rivers dry up, and the other fish in the river die. What happens to a lungfish? It curls up into a ball in the mud at the bottom of the lake or river. Then it begins to breathe with its lungs instead of its gills.

If necessary, a lungfish can live this way for as long as three or four years. When the rains come, the lungfish straightens itself out. Then it begins swimming around in the water again—almost as if nothing had happened!

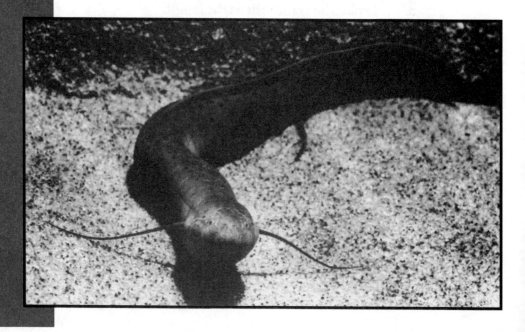

QUESTIONS

1. A lungfish is *different* from other fish because it
   a. can breathe through either lungs or gills.
   b. has gills to breathe through but no lungs.
   c. is not able to breathe air.

2. Most lungfish are found
   a. in the deep waters of the ocean.
   b. in shallow lakes or rivers.
   c. on the shores of rivers.

3. A lungfish can live out of water for
   a. three or four years.
   b. more than six years.
   c. less than four hours.

4. In the story, what happens *first*?
   a. The lungfish curls up into a ball.
   b. The river dries up.
   c. The lungfish breathes through its lungs.

5. You could say that the lungfish is lucky because
   a. it can live either in water or on land.
   b. it need never live in water.
   c. it lives where there is plenty of rain.

LIFE SCIENCE 13

Hide and Seek

Their color and markings help animals to hide.

Look at the picture of the moth. The moth is not easy to see because it is almost the same color as the bark on the tree. The moth's coloring is a kind of *camouflage* (kăm'ə flăzh'). This coloring, or camouflage, helps the moth hide from its enemies.

Many animals have colors and markings that make them hard to see. A white polar bear is hard to see against the snow. In the water, an alligator looks like a floating log. Cheetahs have spots, and zebras have stripes. These markings camouflage the animals' bodies so their enemies cannot see them.

Color also hides insects. Some insects look like sticks and stems. Grasshoppers are as green as the plants they eat. In one forest, an insect is bright pink. It hides on pink flowers.

Look at the animals where you live. List the animals that are hard to see. How many did you find?

QUESTIONS

1. *Camouflage* is important to many animals because
    a. it helps the animals hide from their enemies.
    b. other animals can find them more easily.
    c. it makes them look green.

2. In the water, an alligator
    a. turns green in color.
    b. looks like a floating log.
    c. cannot hide easily.

3. Spots or stripes on animals are also known as
    a. colors.
    b. bark.
    c. markings.

4. If a moth were on a green leaf,
    a. you could not see it easily.
    b. it would be easy to see.
    c. its enemies could not find it.

5. The *main idea* of this story is that
    a. many animals are protected from harm because of their color.
    b. polar bears live in the snow.
    c. some insects are bright pink and live in the forest.

Orcas

Are orcas as alike as peas in a pod?

Peas in a pod may be alike. Orcas in a pod are not. An orca is a black and white whale. Sometimes it is called a killer whale. Many orcas live in family groups called pods. Ten to fifty whales can be in the family. Older females are the pod's leaders. The other whales are all related. They call to each other. They even "speak" the same family slang.

However, the whales do not look alike. Scientists in Canada have photos of 500 different whales. The photos show that each dorsal, or back fin, is different.

Scientists now know that about 300 whales live near the Pacific coast. They hunt for salmon. Two hundred other whales are visitors. They hunt seals, dolphin, and other sea mammals.

Pods of orcas do not hunt all the time. Sometimes the calves play with jellyfish. Or they jump over their mothers. Sometimes orcas leap out of the sea. Then they look like giant balloons. They are not very dangerous, unless you are a fish!

QUESTIONS

1. A family group of orcas is called a _____.

2. An orca is a black and white
    a. jellyfish.
    b. dolphin.
    c. whale.

3. Scientists can tell orcas apart because
    a. baby orcas play with jellyfish.
    b. photos show their dorsal fins are different.
    c. killer whales hunt near the coast.

4. According to the story, these sea mammals must have very good.
    a. hearing.
    b. eyesight.
    c. sense of smell.

5. Killer whales are misnamed because they
    a. do not kill other mammals.
    b. hunt only in family groups.
    c. are known to play with each other.

Ants That Look for Leaves

Did you ever hear of the leaf-cutting ant.

What kind of ant cuts off bits of leaves? The ant is called the leaf-cutting ant, and it lives in parts of Central America and South America.

Leaf-cutting ants build their nests under the ground. Many of them go out to search for leaves. They march along one after the other. The ants stop when they find a tree or bush. Each ant cuts off a piece of leaf. Then the ants march back to their nest carrying the leaf pieces over their heads.

The ants chew up the leaf pieces, but they do not eat them. Instead, they place the wet, chewed-up leaves in the nest. Plants called *fungi* (fŭn′jī′) begin to grow on the leaves. The nest gets no sunlight and is a good growing place for some kinds of fungi. The ants then eat the fungi.

QUESTIONS

1. *Fungi* are
    a. leaves.
    b. plants.
    c. animals.

2. Leaf-cutting ants are found
    a. in parts of Central America and South America.
    b. all over the world.
    c. only in South America.

3. Where do leaf-cutting ants hold the pieces of leaves when they march?
    a. in their mouths
    b. over their heads
    c. on their heads

4. In the story, which of the following things happens first?
    a. The ants put bits of leaves in their nest.
    b. The ants chew up bits of leaves.
    c. The ants eat the fungi.

5. One reason fungi grow on the wet, chewed-up leaves is that
    a. the ants have put the leaves in a dark place.
    b. all fungi grow best in places where there is no light.
    c. the leaves come from a special kind of bush.

A Strange World Under the Sea

Discovered: strange animals under the sea!

It is like another world. The sun never shines there. It is a place far beneath the sea. The ocean water there is very hot and full of *chemicals* (kĕm′ĭ kəlz). It is a vent, a crack in the ocean floor.

For many years, scientists did not know about this undersea place. They had no way to travel safely to places deep under the sea. But now, scientists can make the trip in small underwater ships called *submarines* (sŭb′mə rēnz′).

The scientists have seen many strange water animals in those dark waters. "It was like going to Disneyland," said one scientist. He saw red worms as thick and as long as a jump rope. Giant sea spiders, as big as dinner plates, moved among the rocks. Animals that looked like flowers floated in the water.

How did these strange animals come to be? Why don't they live on land? So far, these are the only animals scientists know of that could live in the deep undersea world.

QUESTIONS

1. A *submarine* is
    a. an animal that lives underwater.
    b. an underwater ship.
    c. a chemical found underwater.

2. In a dark, deep place beneath the sea, scientists saw
    a. strange animals.
    b. a strange ship.
    c. lots of sunlight.

3. The worms the scientists saw
    a. were as big as dinner plates.
    b. looked like flowers.
    c. were thick and long.

4. Why has this strange part of the ocean only just been discovered?
    a. Until now, there was no safe way to get there.
    b. Scientists never wanted to go there before now.
    c. People thought that the animals living there were dangerous.

5. Under which of the headings below would you put the red worms?
    a. Underwater Plants
    b. Strange Land Animals
    c. Unusual Undersea Animals

Think Before You Eat

Why you should watch what you eat

"Drink your milk! Finish your vegetables! Have some fruit instead of eating that candy!"

Most of us have heard these words. But do we know why one food is better for us than another? It is because some foods have more *nutrients* (noo'trē ənts) than others. All plants and animals need nutrients to grow and to stay strong. Your body needs them, too.

You get nutrients from the foods you eat. Different foods have different nutrients, and no one food has all the nutrients your body needs. So you must eat some foods from each of the groups in the picture below. And you must eat them every day. Scientists have arranged food groups into the Daily Food Pyramid. The pyramid shows how much of each kind of food to eat every day. The pyramid helps you to get the nutrients you need. It also tells you what foods can make you sick if you eat too much.

22 LIFE SCIENCE

QUESTIONS

1. People, animals, and plants all need _____ to grow and to stay strong.
    a. fruits
    b. grains
    c. nutrients
2. We get most of the nutrients we need from _____ groups.
    a. two
    b. three
    c. four

Use the Daily Food Pyramid to answer the questions below.

3. What should you eat very little of?
    a. meat
    b. vegetables
    c. fats, oils, sweets
4. What should you eat a lot of?
    a. bread, cereal, rice, pasta
    b. vegetables
    c. milk and cheese
5. What do you think the children playing in the pyramid means?
    a. This pyramid is funny.
    b. Have fun.
    c. Get exercise.

Your Teeth and You

Take care of your teeth and get checkups.

You should visit your dentist at least once a year. The dentist will check your teeth to see if they are growing straight. He or she will check your gums to see if they are firm and pink or light red in color. It is difficult to chew food properly if your teeth are crooked. And soft gums can cause loss of teeth.

The dentist will check to see if your teeth are clean. Are you using *dental floss* (děn'tl flôs')? Are you brushing the right way? The dentist may find a *cavity* (kăv'ĭ tē), or a hole, in your tooth. If so, he or she will clean out the cavity and fill it with a special material.

Eating too many sweets can cause cavities. So the dentist will remind you that drinking milk and eating cheese help make your teeth strong. And eating fruits and vegetables helps make your gums firm and healthy.

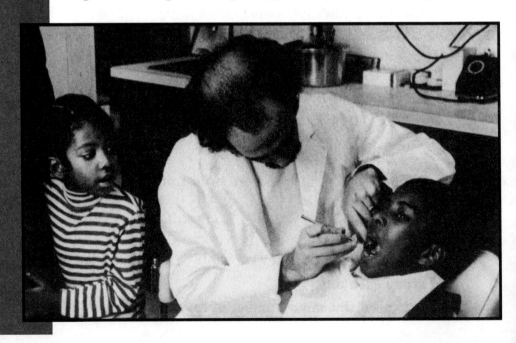

QUESTIONS

1. A *cavity* is a
    a. tooth.
    b. hole.
    c. filling.
2. The dentist should check your teeth
    a. at least once a year.
    b. about once every two years.
    c. every week.
3. According to the story, if your teeth are not growing in straight, you will not
    a. be able to eat any food at all.
    b. have to brush them often.
    c. be able to chew food correctly.
4. In order to have strong teeth and firm gums, you should
    a. eat only fruits and vegetables.
    b. eat the right kinds of food.
    c. drink nothing but milk.
5. The *main idea* of this story is that
    a. the only thing you have to do to have strong teeth is to brush them often.
    b. if you do not eat sweets, you will never have a cavity.
    c. there are many things you must do if you want strong teeth and healthy gums.

Black and White All Over

Do you know who Ling-Ling and Hsing-Hsing were?

In 1972, Ling-Ling and Hsing-Hsing arrived at their new home in the National Zoo in Washington, D.C. Ling-Ling and Hsing-Hsing were pandas (păn' dəz) and were gifts from the people of China to the people of the United States.

Pandas live in the forests and mountain areas of China and Tibet. Ling-Ling and Hsing-Hsing were giant pandas. There are two types of pandas, the giant panda and the red panda. Giant pandas are black and white, have short tails, and grow to be about 3 1/2 to 6 feet tall. Some pandas weigh up to 300 pounds! Because of their size and shape, pandas look very much like bears. They even walk slowly,

just like bears. Some scientists believe that the panda is a member of the bear family, but others disagree.

Pandas like to eat bamboo shoots. Bamboo is a plant that grows in China and Tibet. Workers at the zoo wanted Ling-Ling and Hsing-Hsing to feel at home, so the pandas got lots of bamboo shoots to eat.

QUESTIONS

1. *Panda* is the name given to
    a. an animal that lives in zoos.
    b. an animal found in the forests of China.
    c. all black and white animals.
2. Ling-Ling and Hsing-Hsing
    a. were giant pandas.
    b. were 6 feet tall.
    c. had red fur.
3. The pandas
    a. were born in the zoo.
    b. like to eat bamboo shoots.
    c. have very long tails.
4. Which of the following statements is *true*?
    a. Giant pandas may be part of the bear family.
    b. All pandas are black and white.
    c. The zoo workers did not like taking care of the pandas.
5. The story tells us that
    a. there is more than one type of panda.
    b. Ling-Ling and Hsing-Hsing lived in Tibet.
    c. giant pandas cannot be kept in zoos.

Meet the Cactus

When are bunny's ears green?

When are a bunny's ears green? When does a rat's tail grow in a basket?

Was your answer "When it is a cactus plant"? If so, you are right! Plants that belong to the cactus family are special. Cactus plants can live for weeks without being watered. They store extra water in their roots and stems.

The cactus plant is a green plant. Like other green plants, it uses sunlight to make food. It makes its food in its stem. The body of the cactus is usually covered with spines. These spines can be soft and hairy. Or they can be hard and pointed like a needle. Also, all cactus plants have some type of flower.

There are hundreds of different kinds of cactus plants. Some, like the "living rock" cactus, are small and almost round. Others, like the "bunny ears" cactus, can grow quite tall and wide. And still other cactus plants, like the "rattail," grow long, thin stems that hang down.

QUESTIONS

1. A *cactus* is
    a. an animal.
    b. a plant.
    c. a flower.
2. A cactus plant is *special* because it
    a. is tall and wide.
    b. has spines.
    c. can store water.
3. The cactus plant is like other green plants because it
    a. has scratchy spines that are like needles.
    b. has no flowers.
    c. uses sunlight to make its own food.
4. The food is made in the _____ of the cactus.
    a. stem
    b. spines
    c. roots
5. Cactus plants are good plants to grow in places where
    a. there is a lot of rain.
    b. there is very little rain.
    c. it is always dry.

Sugar Snap Peas

Do you like peas? You will love the "Sugar Snap."

Peas are seeds that we eat as vegetables. Peas are found in long green cases called *pods* (podz̆). These pods, or seed cases, grow on a vine.

Scientists, like Dr. Calvin Lamborn, are always looking for ways to grow better, tastier vegetables. For years, Dr. Lamborn tested many different kinds of seeds. He wanted to grow a better pea. And he succeeded in growing the Sugar Snap pea.

The Sugar Snap is a different kind of pea. You can eat all of it-both the pea and the pod. Usually, a pea pod is tough, and people throw it away. But the Sugar Snap pod is tasty and good to eat. Also, Sugar Snap peas have a wonderful, sweet flavor.

The first seeds for this new pea were sold in 1979. Now, Sugar Snaps are growing in many gardens. People like to pick them from the vine and eat them raw. They say that is when the Sugar Snap tastes the sweetest.

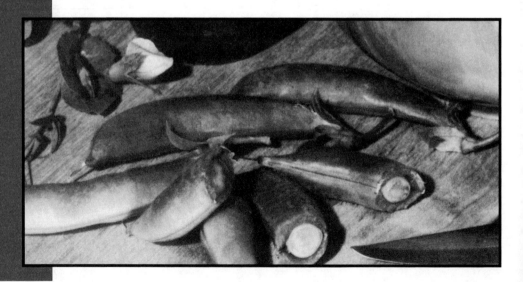

QUESTIONS

1. A *pod* is
    a. a seed case.
    b. a vine.
    c. a kind of pea.

2. Peas are eaten as
    a. candies.
    b. vegetables.
    c. seeds.

3. One reason the Sugar Snap pea is different from most other peas is that
    a. the pod is good to eat.
    b. it does not grow on a vine.
    c. it is not a seed.

4. Why do you think these new peas are called Sugar Snaps?
    a. They have a sweet taste.
    b. They taste sweetest when they are cooked.
    c. They can be used to make sugar.

5. The *main idea* of this story is that
    a. scientists are always looking for ways to grow better foods.
    b. pea pods are not eaten because they are tough and tasteless.
    c. Sugar Snap peas are growing in many gardens.

The Ginkgo Tree

> **The ginkgo is called a "living fossil."**

Ginkgoes (gĭng′ kōz) are a kind of tree. The ginkgo tree has dark, waxy leaves that look something like a fan. The leaves of the ginkgo help to give the tree its full shape.

Ginkgoes have grown on Earth since the time of the dinosaur. We know this because leaves from the ginkgo have been found as *fossils* (fŏs′əlz). Fossils are the remains, or what is left, of plants and animals that lived long ago. They are found in certain kinds of rock. The ginkgo fossils have the same shape as the leaves you see on the ginkgo tree today. The ginkgo's leaves have not changed a bit.

Ginkgoes grow well almost everywhere in North America. They can grow where most other trees cannot. So people often plant ginkgo trees along city streets. Even city smoke and soot do not seem to bother them. No wonder the ginkgo tree has been around for such a long time.

QUESTIONS

1. The remains of plants and animals that lived long ago are called
    a. rocks.
    b. fossils.
    c. leaves.
2. The leaves of the ginkgo
    a. are dark and fan-shaped.
    b. are light green in color.
    c. make the tree look tall and thin.
3. Fossils of ginkgo leaves
    a. prove that the ginkgo is not very old.
    b. have been found in certain kinds of rocks.
    c. do not look very much like ginkgo leaves today.
4. After reading the story, which statement would you say *best* describes the ginkgo?
    a. Ginkgoes have long, pointed leaves.
    b. The ginkgo has been growing on Earth for millions of years.
    c. The ginkgo cannot live in cities where the air is dirty.
5. Which of the following words tells you what a ginkgo is?
    a. fossil
    b. rock
    c. tree

EARTH-SPACE SCIENCE

Jupiter is the largest planet in the solar system. Cameras on Voyager I took photographs of the planet and four of its moons. Later, the photographs were combined to make this picture. It does not show the right size of the moons. But it shows the right positions. Io (upper left) is the closest moon. Europa is in the center. Then comes Ganymede. Finally, you can see Callisto in the right-hand corner. Jupiter has nine other, smaller moons.

WORDS TO KNOW

Spinning Winds
funnel, a cone-shaped object, larger at the top and smaller at the bottom
fierce, violent
tornado, a wind storm of spinning air
whirling, spinning rapidly
tilt, to tip up
predict, to know when something is about to happen

A Black Graveyard for Animals
tar, a thick, sticky brown or black liquid
extinct, no longer in existence
saber, a sword with a slightly curved blade

Moon Secrets
astronauts, persons who fly in a spacecraft
satellite, an object sent up to circle a body in space

A Nosy Neighbor
nosy, curious, prying
crust, the outer surface
canyon, a narrow chasm with steep walls, a gorge
poles, the areas around the top and bottom of a planet

Hurricane Watch
instruments, tools, devices for measuring conditions

Meteorites
alien, from another place
meteorites, stony or metallic objects that reach the Earth's surface; from meteoroids, objects from specks of dust to asteroids weighing many tons that fly about in outer space
Antarctic, the continent and area around the South Pole

Mountains of Ice
glacier, a large mass of ice and snow
steer, to guide

The Alvin
hydrogen bomb, a powerful weapon that causes an explosion large enough to destroy a large city
hydrothermal vents, cracks in the deep ocean floor where escaping gases make water very warm

What Did the Dinosaurs Do?
similar, things that are like one another
appearance, how something looks
image, a picture

Earth-Space Science

EARTH-SPACE SCIENCE 37

Spinning Winds

The power of tornadoes

Don Yancey was parking his car on the side of the highway. He never saw the funnel-shaped cloud coming. As he was about to get out of his car, it got very dark and quiet outside. Then the fierce, spinning winds of a tornado were on top of him. The car began to rock and bounce. Pop! Pop! Pop! The car windows blew out. The winds pulled Yancey out of his car and across the road.

Scientists are studying the power of these whirling winds. They have found that tornado winds can spin at 100 miles an hour and more. Recently a tornado in Oklahoma had winds of 318 miles an hour.

Tornadoes form during big wind storms. As two layers  of wind blow in different directions, air trapped between the layers begins to spin. Strong winds blowing up tilt the spinning air on its end. Other winds blowing down through the spinning air produce a tornado.

Learning about tornadoes helps scientists to predict them and warn us before they come.

QUESTIONS

1. A tornado is a
    a. rain storm.
    b. twisting, whirling wind.
    c. thunderstorm.

2. Tornadoes form
    a. during wind storms.
    b. over highways.
    c. in the winter.

3. The highest known speed of tornado winds is
    a. 100 miles an hour.
    b. 200 miles an hour.
    c. 318 miles and hour.

4. If it suddenly gets very dark and quiet in the middle of the day,
    a. the sun has set.
    b. a tornado could be very near.
    c. it is going to snow.

5. Learning about tornadoes is good because
    a. we can get a warning before they hit.
    b. tornadoes are a kind of weather.
    c. people can make movies about them.

A Black Graveyard for Animals

What are the La Brea Tar Pits?

Over a hundred years ago, Henry Hancock owned a ranch in California. There were sticky, black tar pits on the land. They smelled like oil and gas. The pits were named the La Brea Tar Pits.

Mr. Hancock decided to sell the tar to some road builders. But the workers were surprised when they began to dig. The tar was full of old bones! Many of the bones were of animals that were *extinct* (ĭk stĭngkt′), or had disappeared from Earth, a long time ago. Some of these extinct animals were a giant bird, a "short-faced" bear, a tiny horse, and a saber-toothed tiger.

One day, someone discovered how the animals had been trapped in the pits. A man was watching the shiny pools of tar. He saw a duck try to get a drink from a tar pit. But it got stuck in the bubbling tar and cried out. Hungry animals heard the duck and got stuck trying to catch it. Over thousands of years, many animals died this way.

QUESTIONS

1. Something that disappeared from Earth a long time ago is
    a. trapped.
    b. extinct.
    c. discovered.
2. What did the tar pits smell like?
    a. oil and gas
    b. drinking water
    c. ranch animals
3. According to the story, tar can be used in making
    a. roads.
    b. oil and gas.
    c. paint.
4. What did the animals sometimes think the tar pits looked like?
    a. something good to eat
    b. old bones
    c. drinking water
5. Why did other animals get stuck after the duck got stuck?
    a. They were after a meal.
    b. They tried to help the duck.
    c. They were afraid.

Moon Secrets

What do we know about the moon?

Was there ever life on the moon? Is the moon like Earth? How old is the moon? The United States sent *astronauts* to find out. An astronaut is a person who travels in space. From 1969 to 1972 astronauts made ten moon landings and brought back rocks from each trip.

Scientists learned a lot from the moon rocks. There was probably never life on the moon. The moon formed the same time as Earth. Some scientists think the moon broke off from the Earth. That was long ago!

Moon rocks show that the Earth and the moon changed in the same ways at first. Then about 3 billion years ago, they began to change in different ways. We know when the changes happened, but not why.

Scientists still want to know more about the moon. In 1999 they crashed a *satellite* into the moon. This was to check for water vapor. No vapor was seen. But scientists still think there may be water there.

QUESTIONS

1. An *astronaut* is
    a. a person who travels into space.
    b. one kind of rock found on the moon.
    c. something that lives on the moon.

2. The moon rocks show that there
    a. is life on the moon now.
    b. was probably never any life on the moon.
    c. was life on the moon over 3 billion years ago.

3. The moon rocks show us _____ the moon and Earth were formed.
    a. when
    b. how
    c. why

4. The secrets we have learned from the moon's rocks tell us
    a. what is happening on Earth now.
    b. what has happened on the moon in the past.
    c. what will happen on the moon in years to come.

5. According to the story, which of the following is *true*?
    a. There are many things yet to be learned about the moon.
    b. By studying the moon rocks, scientists have learned all the moon's secrets.
    c. Studying the moon rocks helped scientists learn why the moon changed.

A Nosy Neighbor

Where is your water, Mars?

Mars is a neighbor of Earth. Some day we hope to visit this neighbor. Before we do, scientists want to know more about Mars. They are especially nosy about the Red Planet's crust.

Photos show giant red clouds. They seem to be dust storms, not rainstorms. Other photos show hundreds of rocks. Some big, some small. There are also sand dunes, red dust, mountains, and deep canyons. It looks as if huge rivers had once moved the rocks and piled the sand. But the photos show no water.

Scientists want to know what happened to the water. When was there ever water? Are the large white areas near the poles ice? Could Mars explorers melt the ice for water? Was there ever enough water for life on Mars? If so, what happened to those living things? The nosy scientists will keep searching until their questions are answered.

QUESTIONS

1. The outside of a planet is called the _____.
2. Mars is called the Red Planet because it has red
    a. sunsets.
    b. rain.
    c. storm clouds.
3. According to the story, scientists study Mars' crust by using
    a. X-ray machines.
    b. photographs.
    c. ice machines.
4. Ice on Mars means there
    a. was once water.
    b. are many living things.
    c. will still be water for settlers.
5. Which of the following statements is *true*?
    a. There is no life on Mars.
    b. Scientists agree that we will never be able to land on Mars.
    c. There are many questions still to be answered about Mars.

Hurricane Watch

Pilots fly right into a hurricane. Why?

Hurricanes (hûr′ĭ kānz′) are big storms that start over water and bring strong winds. Hurricane winds can blow at 72 miles per hour or more and must be watched.

Who watches a hurricane? Hurricane hunters do. They are weather scientists who fly into a hurricane in an airplane.

Inside the plane, there are many instruments that the hurricane hunters use to find out things about the hurricane. What kind of information are these weather scientists looking for?

For one thing, hurricane hunters need to know in which direction the hurricane is moving. So they use the instruments to track the storm. If it starts moving toward land, they send out warnings. Then people can move out of the way before the hurricane hits. These hurricane hunters help save many lives.

QUESTIONS

1. A *hurricane* is a
   a. big storm with strong winds.
   b. kind of airplane.
   c. wind instrument.

2. Hurricanes start
   a. in the mountains.
   b. on land.
   c. over water.

3. Which of the following statements is *not* true?
   a. Hurricane hunters are weather scientists.
   b. Hurricane hunters use instruments to track storms.
   c. Hurricane hunters do most of their work on the ground.

4. According to the story, hurricane hunters watch a storm from
   a. an airplane.
   b. a tracking station.
   c. a boat.

5. Hurricane hunters are important because they
   a. are able to stop storms.
   b. warn people when storms are coming.
   c. know how to fly airplanes.

Meteorites

Are there alien rocks on Earth?

Some people said that rocks fell from the sky. No one believed the stories. Where could the rocks come from? How could you tell?

Slowly, scientists began to collect some facts. Sometimes, people saw groups of lights fall from the sky. The "lights" fell nearby. The next day people found strange rocks on the ground. Often the alien rocks were black and heavy. They had strange shapes and were unlike nearby rocks.

Finally, scientists were sure these rocks were not from Earth. They are part of meteors. Usually meteors burn up before hitting Earth. But a few pieces can reach the ground. These pieces are called meteorites. By studying them, scientist can learn about space.

Today, scientists hunt for meteorites everywhere. One place is Antarctica. More than 20,000 have been found there. But hunting for rocks near the South Pole is hard work. So scientists built a robot, called Nomad, to help them search.

QUESTIONS

1. Another word for alien is _____.
2. Pieces of meteors found on Earth are called _____.
3. According to the story, "lights" falling from the sky are really
    a. burning rocks.
    b. alien invaders.
    c. burning robots.
4. It is easy to find meteorites on Antarctica because they
    a. are still very hot.
    b. fall on the South Pole.
    c. show up against the ice.
5. Which tools on the robot Nomad will help show that a rock is a meteorite?
    a. Antenna and science manipulation arm
    b. Laser range finder and transforming chassis
    c. Spectrometer and high-resolution camera

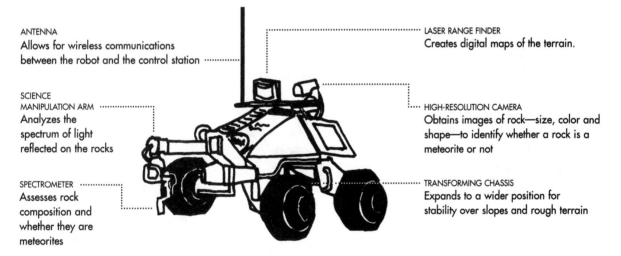

ANTENNA
Allows for wireless communications between the robot and the control station

SCIENCE MANIPULATION ARM
Analyzes the spectrum of light reflected on the rocks

SPECTROMETER
Assesses rock composition and whether they are meteorites

LASER RANGE FINDER
Creates digital maps of the terrain.

HIGH-RESOLUTION CAMERA
Obtains images of rock—size, color and shape—to identify whether a rock is a meteorite or not

TRANSFORMING CHASSIS
Expands to a wider position for stability over slopes and rough terrain

Mountains of Ice

Icebergs can be dangerous.

It is April in the North Atlantic Ocean. Suddenly, there is a sound like thunder. A giant block of ice breaks off from a *glacier* (glā′shər). An iceberg is born.

Berg is a German word for "mountain." An iceberg is like a mountain of ice rising above the water. But the part of the iceberg that is above the water is only a small part of the whole thing. Most of the iceberg is underwater and cannot be seen. The bottom of a ship could be torn open if it hit this hidden part of an iceberg.

The United States Coast Guard uses planes and ships to watch the waters where icebergs are usually found. The Coast Guard also keeps track of the iceberg when it is spotted. Then the Coast Guard can warn ships to steer around it. These iceberg watchers have helped save many ships and many lives.

QUESTIONS

1. The German word for "mountain" is
    a. kilometer.
    b. thunder.
    c. berg.
2. Most of the iceberg is usually
    a. above the water.
    b. made of salt.
    c. under the water.
3. An iceberg is dangerous because
    a. most of it is hidden.
    b. it breaks off from a glacier.
    c. it looks like a mountain.
4. When ships see an iceberg, they
    a. must return to shore.
    b. go around it.
    c. push it away.
5. The United States Coast Guard helps save ships and lives by
    a. tracking icebergs and warning ships.
    b. melting the tops of icebergs.
    c. steering icebergs out of the ships' way.

The Alvin

A tiny ship takes scientists into the sea.

The *Alvin* is a deep-diving *submersible*. A submersible is used to travel underwater. Scientists use the *Alvin* to study the deep sea. In two hours, the *Alvin* can go deeper than one mile. The water temperature is near freezing at this level. But inside the *Alvin* it is warm and comfortable. A pilot will steer the ship around the ocean floor. The two scientists inside can use the *Alvin*'s two robotic arms to move things outside the ship and collect samples. They can see outside through the three windows. The *Alvin* has three video cameras. It has two cameras that take still pictures. Because there is no light in the deep sea, the Alvin has headlights.

The *Alvin* was built in 1964 and has been used on more than 3,500 dives. In 1966, it located a hydrogen bomb that had been lost off a ship. It explored hydrothermal vents discovered in the Galapagos. The *Alvin* also helped find the *Titanic*.

QUESTIONS

1. If something is *submersible*, it
    a. can take pictures.
    b. can go underwater.
    c. has more than two windows.

2. A scientist would use the *Alvin* mainly to
    a. study the deep parts of the sea.
    b. travel from one place to another.
    c. learn how to dive in the deep sea.

3. When the *Alvin* is one mile deep, the
    a. people on board are uncomfortable.
    b. water temperature is about freezing.
    c. inside of the ship becomes very cold.

4. Why does the *Alvin* have headlights?
    a. So people on the ship can see each other.
    b. So scientists can look at the sea floor.
    c. So the pilot can see which button to push.

5. According to the story, one of the *Alvin*'s jobs was to
    a. help whales find their way south in the winter.
    b. locate schools of fish.
    c. help find the wreck of the *Titanic*.

What Did Dinosaurs Do?

What do fossils tell us?

Dinosaurs were animals that lived long ago. Their bones have become fossils. A fossil is what is left of plants or animals that lived long ago. From fossils scientists have put together dinosaur skeletons. We now know what many dinosaurs looked like.

But what did they do? Their footprints teach us that dinosaurs moved in large groups, or herds, as some animals do today. They may have traveled long ways to find food or water. Findings of groups of nests show us that some dinosaurs may have lived in family groups.

Scientists also study animals that are similar in size or looks to the dinosaurs. In this way they try to learn more about dinosaurs. How fast dinosaurs could run? What did they eat? Could Apotosaurus, a type of dinosaur, stand on its hind legs? Today, with computers scientists create images. They can then make them move just like a dinosaur might really have moved.

54 EARTH-SPACE SCIENCE

QUESTIONS

1. Many plants and animals that lived long ago are now found as
    a. fossils.
    b. rocks.
    c. sand.
2. Scientists believe that dinosaurs may have traveled in herds because they
    a. saw the movie *Jurassic Park*.
    b. have found tracks that show them this.
    c. were told by very old people.
3. Scientists figured out what many dinosaurs looked like by
    a. putting together skeletons.
    b. looking at drawings on cave walls.
    c. studying birds.
4. How are scientists figuring out how dinosaurs could move?
    a. by digging for their bones
    b. by studying their nests
    c. by creating images on computers
5. When scientists found groups of fossilized dinosaur nests they realized that
    a. dinosaurs may have lived in family groups.
    b. dinosaurs may have been fast runners.
    c. dinosaurs may have been meat-eaters.

PHYSICAL SCIENCE

Up, up, and away! These people are going for a ride in a hot-air balloon. The air inside the balloon is being warmed by the small heater under the balloon's opening. When the balloon is full of warm air, it will rise. It will lift the basket and the people high into the air. How do you think they will get down?

WORDS TO KNOW

Hang Gliding
soar, fly high in the air
control, means of directing

Hot-Air Balloon
discovery, a first finding
straw, dry stalks of grain from which the grain has been harvested
nylon, a human-made fiber of great strength and flexibility
breathtaking, very exciting, thrilling

Maglev
magnet, a piece of metal that attracts iron
metal, a hard, strong, shiny sort of material
force, strength or power that does work, such as moving something
magnetism, the strength a magnet has
engineers, people who write directions for building things

How a Battery Works
cells, small enclosed spaces
case, container
rod, a straight stick or bar
paste, any soft, moist, smooth substance

The Dangerous Rays of the Sun
ultraviolet, the sun's rays that cause sunburn
protection, to shield something from being hurt
sunscreen, a liquid or cream used to protect the skin from the ultraviolet rays of the sun
brim, the edge, or rim, of a hat that sticks out from the hat

Lasers in Medicine
focused, to bring light rays to shine on a certain point
weird, odd, strange, unusual

Physical Science

Hang Gliding

Would you like to ride the winds in a large kite?

Did you know that people can soar and float in the wind? Well, they can. They use hang gliders. Hang gliders are like large kites.

The person who rides a hang glider is called a pilot. The pilot holds on to the hang glider and runs down a hill. The hang glider lifts off into the air. The pilot rides in a seat under the glider.

In the air, the pilot uses a control stick to steer the glider up, down, and from side to side. The pilot tries to find a thermal. A *thermal* (thûr′məl) is a wave of rising hot air. The rising hot air keeps the glider up. Mountain areas are good places to find thermals.

Hawaii is a good place to go hang gliding. The winds are strong, and there are good thermals. One pilot was able to hang glide in the air for over 12 hours!

QUESTIONS

1. A wave of rising hot air is called
    a. a glider.
    b. a thermal.
    c. a kite.
2. In this story, a pilot is someone who
    a. rides a hang glider.
    b. makes large kites.
    c. lives in Hawaii.
3. A control stick is used to
    a. keep a hang glider on the ground.
    b. hold the pilot in his or her seat.
    c. steer a hang glider up, down, and sideways.
4. Why is a thermal important in hang gliding?
    a. It keeps the glider up in the air.
    b. It makes the air warm.
    c. The pilot uses it to hang on to the glider.
5. If you wanted to hang glide, you would go to a place where
    a. the air is cold.
    b. the winds are light.
    c. the thermals are good.

Hot-Air Balloon

How does a hot-air balloon work?

Many years ago, in 1783, two Frenchmen found out that a paper bag filled with hot air will float. From this discovery came the idea that people might be able to travel by hot-air balloons. People simply made giant paper bags and got into a basket hooked to the bottom of the bag. Next, they filled the bag with hot air, and up they went! People used to burn straw to keep the air inside the balloon hot.

Today, hot-air ballooning is different from what it was in those early times. Now, the balloons are made of nylon, not paper. Today, balloonists use a gas burner to keep the air hot. But ballooning is still just as exciting.

Once off the ground, balloonists are in for a breathtaking ride. Except for the sound of the gas burner, the ride is quiet. No wind is felt, since the balloon travels with the wind. Finally, the balloonist turns down the gas burner. The air in the balloon cools, and the balloon floats back to the ground.

QUESTIONS

1. A balloon is a
    a. special kind of gas.
    b. bag filled with air.
    c. basket hooked to a bag.
2. No wind is felt on a balloon ride, because
    a. the balloon travels with the wind.
    b. the people are inside the balloon.
    c. the distance traveled is not great.
3. To come back to the ground, the balloonist
    a. throws the gas burner out of the balloon.
    b. lets air out of the balloon.
    c. turns down the gas burner.
4. The balloons in this story work because
    a. people steer them.
    b. hot air rises.
    c. cold air is light.
5. Traveling in a balloon is probably most like
    a. flying in an airplane.
    b. riding in a car.
    c. floating on a cloud.

Lasers in Medicine

What is a laser?

The word laser stands for "light amplification by stimulated emission of radiation." A laser is simply a beam of light. It is focused so much that it seems to have edges. Because a laser beam stays so sharp, it's very useful in many ways. A laser beam can scan groceries to check prices. It can send information through phone cables.

One of the most important ways lasers are used is to heal people. Because a laser beam is so focused, it can be very hot. It isn't dangerous to patients-only to tumors. Tumors are weird lumps of cells that often need to be removed. A laser can burn up the tumor in a second without hurting anything else. A laser beam can be focused to the size of a pinpoint, so it can also be used in eye surgery. Laser beams don't leave scars, either. They can even be used to remove tattoos!

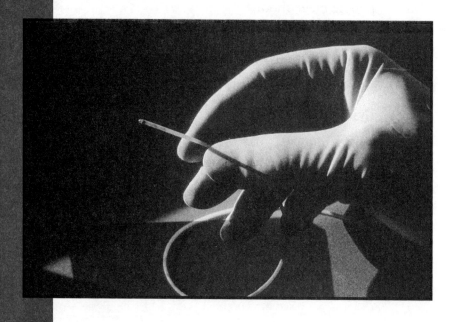

QUESTIONS

1. A laser beam is
   a. a wooden or steel beam that supports a building.
   b. a beam of light.
   c. a flashlight.
2. Because a laser beam is so focused it
   a. is very hot.
   b. can be dangerous to patients.
   c. causes tumors.
3. The main idea of paragraph two is
   a. laser beams don't leave scars.
   b. lasers are used to heal people.
   c. lasers are useful in eye surgery.
4. Lasers are important in medicine because
   a. they can do many operations better than surgery.
   b. they don't hurt people.
   c. they can remove tattoos.

What Do You Think?

Noise is all around us.

When you see the word *dangerous* (dān′jər əs), what do you think about? Fire? Guns? Broken glass? How about noise? Can noise be dangerous? Can it hurt you?

Most people would say that noise is loud, unwanted sound. They think of the roar of traffic, the scream of jet planes, and the howl of a sick animal. But what about rock music? Is the sound of music noise?

Sounds can be measured to find out how loud they are. Your hearing can also be measured. Tests were done on people who have worked with very loud machines for a long time. The tests showed that some of these people had a loss of hearing. People who race cars, fly jet planes, or play in rock bands may also have hearing problems.

Tests also show other changes in people who must be in loud, noisy places. These changes can take place in the way people breathe or see or think or move or feel. Can noise be *dangerous*? What do you think?

QUESTIONS

1. To test a sound, we must
    a. change it.
    b. measure it.
    c. time it.

2. Tests show a loss of hearing in people who
    a. measure loud sounds.
    b. play rock music on their radios.
    c. work with very loud machines for a long time.

3. Which of the following can be dangerous?
    a. all sounds
    b. most sounds
    c. very loud sounds

4. Another name for this story could be
    a. "Sounds Can Be Measured"
    b. "Noise Can Be Dangerous"
    c. "How to Test Your Hearing"

5. People who fly jet planes may have hearing problems. One reason may be that jets
    a. fly too fast.
    b. fly too high above the ground.
    c. make a loud noise on takeoff and landing.

How a Battery Works

Many new games and tools work on batteries.

A *battery* (băt′ə rē) is made up of one or more cells. The cells are used to make electricity. One kind of battery is the D-cell battery. D-cell batteries are small. So they can be used in such things as flashlights, clocks, electric games, and tape recorders.

One kind of D-cell is made up of a zinc case. There is a rod of black material called carbon in the middle of the case. The rest of the cell is filled with a *chemical* (kĕm′ĭ kəl) paste. The zinc and the carbon have to be connected in some way for the cells in the battery to make electricity. Then electricity flows and the toy or flashlight works.

The D-cell battery stops working when the zinc wears out or the chemical paste is used up. Your toy or tool will stop working when this happens. Then it is time to buy a new battery.

QUESTIONS

1. A *battery* is _____ used to produce electricity.
    a. a kind of electric wire
    b. made up of one or more cells
    c. a special chemical
2. The D-cell is
    a. used mainly in cars.
    b. used only in flashlights.
    c. a small battery.
3. The black material in the center of a D-cell may be
    a. carbon.
    b. zinc.
    c. paste.
4. For the cells in a battery to make electricity, there must be a connection between
    a. the zinc and the carbon.
    b. the paste and the battery.
    c. the carbon and the toy or tool.
5. If a battery-powered game stopped working, it could be that
    a. the rod was made of carbon.
    b. there were chemicals in the paste.
    c. the zinc wore out.

Dangerous Rays

How much is too much?

The people in the picture may look funny. They are really very smart. They know they need to shade their skin and eyes from the dangerous ultraviolet rays of the summer sun.

You can find out when these rays are most dangerous. Look for your shadow. Look early in the morning and late in the afternoon. Your shadow will be very tall. This is when you need the least protection. Between 10:00am and 3:00pm your shadow will be very short. This is when you need the most protection. Put on sunglasses, sunscreen lotion, and a hat with a wide brim. These will protect your eyes, face, and neck. Long pants and a shirt with long sleeves will protect your body.

Most people can stay in the summer sun for about half an hour before they start to burn. People with light skin burn faster than people with dark skin. But we all need protection from the dangerous ultraviolet rays of the summer sun.

QUESTIONS

1. Ultraviolet rays come from
    a. dirty air.
    b. TV.
    c. the sun.
2. Most people will start to burn after they have been in the summer sun
    a. for several hours.
    b. for several days.
    c. for about half an hour.
3. You will know you need lots of protection from the sun when
    a. your shadow is very tall in the morning.
    b. your shadow is very short after lunch.
    c. your shadow is very tall in the afternoon.
4. To protect your nose from the sun's ultraviolet rays you need
    a. sunglasses and sunscreen lotion.
    b. long pants and a shirt with long sleeves.
    c. sunscreen lotion and a hat with a wide brim.
5. To protect your body from the sun's ultraviolet rays you need
    a. sunglasses and sunscreen lotion.
    b. long pants and a shirt with long sleeves.
    c. sunscreen lotion and a hat with a wide brim.

The Flying Train

The Maglev goes very fast, just inches above the track.

Some call the Maglev a plane that flies very low. But Maglev is really a super fast train. It rides on air just above its track. It can go 300 miles an hour. It could travel the 224 miles between Washington, D.C. and New York City in less than an hour. And the Maglev uses less energy than other trains.

To know how the Maglev works, get a small magnet. Magnets will attract iron metal. Now put a pin or a needle near the magnet. The magnet pulls the metal toward it. That pulling force is called magnetism. It is magnetism that makes the Maglev ride on air.

The Maglev was invented by American engineers at a college in Boston. But to ride a Maglev you have to go to Japan or Germany. In Japan, the Maglev is called the bullet train. In Germany it is called the Transrapid train. Soon Maglevs may be zooming over tracks in this country.

QUESTIONS

1. An object that a attracts iron is called
    a. magnetism.
    b. energy.
    c. a magnet.

2. The force that makes the Maglev train go is called
    a. air.
    b. magnetism.
    c. a magnet.

3. The Maglev is
    a. a plane.
    b. a train.
    c. a magnet.

4. Maglevs are good because
    a. they are fast and use less energy.
    b. they are big.
    c. they carry a lot of people.

5. If a metal is not pulled by a magnet
    a. it has no magnetism.
    b. it is not iron.
    c. it has no energy.

The Uses of Petroleum

Why is petroleum so important to us?

Do you read at night, or watch TV, or listen to the radio? If you do, you are using some form of *electrical power* (ĭ lĕk′ trĭ kəl pou′ər).

Much of the electrical power in this country is made by *burning* petroleum (pə trō′ lē əm).

Petroleum means "rock oil," and it comes from deep in the ground. Cars, trucks, planes, and buses burn some form of petroleum as *fuel* (fyōō′əl). Fuel is a material that is burned to make heat or power. The light, heat, and power used in many homes come from burning petroleum.

We also use petroleum for many other things. Petroleum is used to make paint, tires, plastics, safety glass, face creams, movie film, shampoo, and hundreds of other things that we all use each day.

Perhaps, in the future, scientists will find another material to take the place of petroleum. Until then, petroleum is very important to us.

QUESTIONS

1. The word *petroleum* means
   a. power.
   b. rock oil.
   c. fuel oil.

2. We get power from petroleum by
   a. burning it.
   b. storing it.
   c. digging for it.

3. Cars, trucks, planes, and buses all burn some form of petroleum as
   a. oil.
   b. fuel.
   c. heat.

4. What is the *main idea* of this story?
   a. Some of the electrical power used in this country comes from burning petroleum.
   b. Cars burn some form of petroleum as fuel.
   c. Without petroleum, we would have to change the way we live.

5. "The Uses of Petroleum" might also have been called
   a. "Making Paint and Plastics from Petroleum"
   b. "How We Heat Our Homes with Petroleum"
   c. "Petroleum: A Very Special Kind of Material"

ENVIRONMENTAL SCIENCE

The zebras are having a drink to satisfy their thirst. They remind us how every living creature on Earth needs water to live. Without water, Earth would be as empty and barren as the moon. One of the great challenges we face in the 21st Century is to protect and to restore our precious water supplies ... from ponds and lakes and rivers to the great oceans.

WORDS TO KNOW

Saving the Bald Eagle
endangered, a type of animal or plant in danger of no longer existing
reason, a cause
hatch, the young coming out of eggs

Solar Heat
solar, having to do with the sun
collector, a device that gathers

Fuel for Tomorrow
energy, power for doing work

Acid Rain
acid, any chemical compound that reacts with a base to form a salt
pesticides, chemicals used to kill insects or other plant pests

We Need a Pollution Solution
polluted, contains things that are harmful to plants, animals, and people
pesticides, chemicals used to kill insects or other plant pests
runoff, rain water not absorbed by the soil

Helping Mother Nature
compost, a mixture of decaying grass, leaves, etc.
crushed, broken into small pieces, **scraps**, small bits of leftover food

Charged-up Cars
petroleum, a natural, yellowish-to-black liquid that will burn, usually found beneath the Earth's surface. Also called "crude oil."
fuel, anything used to produce energy
batteries, devices for storing and generating electricity
charged up, to fill up

Bottle Bills Reduce Waste
litter, anything carelessly thrown away
waste, useless or worthless things
beverage, anything to drink
recycle, to use again
deposit, money

Mountain Gorillas and Dian Fossey
endangered, a type of animal or plant in danger of no longer existing
extinction, no longer existing

Environmental Science

ENVIRONMENTAL SCIENCE **79**

The Bald Eagle Is Soaring Again

People are helping a big bird.

The bald eagle is a big, beautiful bird. It has a snow white head and large wings that can stretch more than six feet.

Bald eagles live near lakes, rivers, and sea coasts. In the 1960s, bird watchers found that fewer and fewer eagle eggs were hatching. The bald eagle was becoming an endangered animal. It was in danger of disappearing.

Why were these birds endangered? One reason is that people were cutting down many trees where eagles nested, or lived. Also, people were using the chemical DDT to kill insects. The DDT was carried by the wind to lakes and rivers. When the mother eagle ate the fish from these waters, she also ate the DDT. This made the shells of her eggs so thin that they broke under her in the nest. Because of this, the eggs did not hatch.

Today, there is a law against using DDT. Other laws set aside places where eagles can safely build their nests and hunt for food. There are now more bald eagles in the United States than there were in the 1960s.

QUESTIONS

1. When fewer and fewer eagle eggs hatched each year, bird watchers knew that the bald eagle had become _____ animal.
    a. an endangered
    b. a dangerous
    c. a plentiful
2. The eagles lost their nests when
    a. the baby eagles were hatched.
    b. people passed laws against eagles' living in trees.
    c. people began to cut down trees where eagles lived.
3. Fewer eagle eggs were hatching because
    a. mother eagles were not laying as many eggs.
    b. DDT was killing the baby eagles.
    c. DDT made the shell of the eagles' eggs very thin.
4. The law against using DDT is a good law because
    a. DDT kills bugs and insects that are harmful.
    b. DDT endangers birds, fish, and other animals.
    c. DDT made the eagles bald.
5. The bald eagle got its name because
    a. from a distance, the snow-white feathers make its head look bald.
    b. the DDT made the feathers on its head fall off.
    c. eagles do not have feathers on their heads.

Solar Heat

The future of solar heating looks very bright.

The heat we get from the sun is called *solar* (sō′lər) heat. Solar heat can be used to heat water and the inside of houses. The picture below shows one kind of solar-heated house.

First, panels are placed on the roof of the house. Each panel is a *collector* (kə lĕk′tər). The collector traps the sun's heat. Water flows through the panels and is heated. Then the heated water flows through pipes in the house and heats the house.

Cloud cover can cause problems. When the sun is hidden by clouds, the panels cannot collect heat. If too many days go by with no sunshine, there may be no heat or hot water. For this reason, people who use solar heat usually have a back-up system, or another way of heating their house.

QUESTIONS

1. The word *solar* means of or by the
    a. air.
    b. sun.
    c. heat.
2. In the story, the panels on the roof of the house
    a. are put on last.
    b. trap water.
    c. collect the sun's heat.
3. As the water flows through the panels, it
    a. is trapped inside them.
    b. cools off the house.
    c. is heated by the sun.
4. How does cloud cover affect a solar-heating system?
    a. The solar collectors cannot trap the sun's heat.
    b. The water will not flow through the pipes in the house.
    c. The solar panels are not able to trap hot water.
5. Which of the following things happens *first*?
    a. Water flows through the panels.
    b. The house is heated.
    c. The collector traps solar heat.

ENVIRONMENTAL SCIENCE

Fuel for Tomorrow

Can cars run without gasoline?

A fuel (fyoo'əl) is a substance that can be burned to make energy. Today, gasoline is used as a fuel for most cars. Tomorrow, other kinds of fuels will be used.

One idea for fuel is to use natural gas. Natural gas is made up of mostly methane (meth' ān). Natural gas is used for cooking, making hot water, and for heating homes. In Egypt, there are many cars and buses that run on natural gas. In that country, natural gas costs less than gasoline. Natural gas burns cleaner than gasoline. The air stays cleaner.

There are problems, however. Methane gas takes up a lot of space so cars need big fuel tanks. Safe, clean natural gas will be another fuel for the cars of tomorrow.

QUESTIONS

1. The word *fuel* is used for anything that
    a. can be burned to make energy.
    b. is put into a tank.
    c. is expensive to burn.

2. According to the story, natural gas is _____ than gasoline.
    a. dirtier.
    b. more expensive.
    c. cleaner.

3. According to the story, one of the problems with using natural gas for fuel is that the gas
    a. takes up a lot of space.
    b. is hard to find.
    c. does not burn very easily.

4. In the story, why do you think some people in Egypt drive cars that use natural gas?
    a. Natural gas is cheaper than gasoline.
    b. Natural gas is more powerful than gasoline.
    c. Natural gas costs the same as gasoline.

5. Suppose your car used methane gas. Which one of the following statements is the *best* reason for getting a larger fuel tank?
    a. You would not have to refuel so often.
    b. The extra methane will help make the car heavier.
    c. You would burn less of the methane.

Acid Rain

Harmful Rain?

Some years ago, people saw that the paint on their cars was being damaged. At first, no one could figure out why. Now we know that acid rain can cause harm. It harms the paint on cars and even the outside of buildings. When we burn coal and other fuels, chemicals are released into the air. These chemicals pollute the air. Such air is bad for us to breathe. When they combine with rain, they can damage many surfaces. Over long periods, acid rain eats away at paint, metal, and stone. Acid rain harms forests because it can stunt the growth of trees. Acid rain falling over a lake can make the water too acidic for the fish that live in it.

In 1990, Congress passed the Clean Air Act. Thanks to this, fewer harmful chemicals are released into the air. Rainfall has become less acidic. But acid rain is still a problem our country cannot ignore.

QUESTIONS

1. Rain that contains chemicals released into the air when fuels are burned is called
    a. lake water.
    b. ground fog.
    c. acid rain.

2. According to the story, acid rain
    a. is no longer a problem since Congress passed the Clean Air Act.
    b. is still a problem.
    c. never was a problem; it was just that the cars had bad paint jobs.

3. What happens first?
    a. Chemicals in the air combine with rain.
    b. Chemicals are released into the air when fuel is burned.
    c. Surfaces are damaged by rainfall which contains certain chemicals.

4. Acid rain has been observed to cause damage to
    a. paint on cars, and to forests, lakes, and buildings.
    b. only ponds but not large lakes.
    c. only human-made objects.

5. Acid rain causes damage
    a. within a few minutes.
    b. within a week.
    c. over a long period of time.

We Need a Pollution Solution

What causes water pollution?

Would you like to drink sewage? How about cleaning fluid? No! Those are some things in *polluted* water. Polluted water can kill fish and birds. It makes us sick.

How does this work? Drinking water comes from sources like rivers and lakes. Factories may use poisons, like mercury and lead, to make things. Factory wastewater then gets into rivers and lakes.

Water is also polluted by acid rain. This takes place when polluted air mixes with rain. The rain then falls into rivers and lakes.

Some farms also pollute. Rain washes *pesticides*, used to kill bugs, into streams. This is called runoff. Runoff has dirt in it, too.

Oil can also get into water. For example, ships may leak oil. Many birds and fish die each year from oil spills.

Polluted water may look clean. So, watch out where you swim. Know the source of the water you drink. Most of all, try to find pollution solutions!

88 ENVIRONMENTAL SCIENCE

QUESTIONS

1. Polluted water is water that
    a. it is okay to drink, if it looks clear.
    b. has harmful things in it.
    c. is in all our rivers.

2. How does wastewater pollute?
    a. Factories sometimes use poisons and these get into nearby water.
    b. Acid rain is the same as wastewater.
    c. Wastewater comes from oil spills.

3. How do farms sometimes pollute our water?
    a. Farmers dump oil in our drinking water.
    b. Farmers use pesticides, and it runs off plants into streams.
    c. Farmers own the factories that pollute.

4. Which is *true*, based on the story?
    a. All farms pollute.
    b. All factories pollute.
    c. Some factories and farms pollute.

5. If there was a lake near a factory or a farm and the lake looked clean, you should...
    a. swim in it.
    b. drink out of the lake.
    c. first find out if the water is safe.

Helping Mother Nature

How can you help mother nature?

Plants need good, rich soil to help them grow. Mother Nature makes some soil rich from things that used to be alive. Then she mixes in small amounts of minerals. This good, rich soil is called *compost*.

You can help Mother Nature. You can make your own compost. Mix equal amounts of garden waste and kitchen scraps. This will make compost that doesn't smell.

Dry grass cuttings, leaves, and weeds are some of the garden waste you can use. Use coffee grounds, tea bags, crushed eggshells, and fruit and vegetable scraps from your kitchen.

Your compost won't take too much room in your garden. But it will take several months to be ready to use. You will know it is ready when it has turned into dark, fluffy soil.

Mix your compost deep into your flower or vegetable beds. Soon, you will see how much better your garden looks. For just like Mother Nature, you will have helped to make things grow.

QUESTIONS

1. The word *compost* means
    a. good, rich soil.
    b. a pile of kitchen scraps.
    c. a pile of garden waste.

2. To keep a compost pile from smelling, you will need
    a. lots of water.
    b. equal amounts of garden waste mixed with kitchen scraps
    c. lots of light.

3. Your compost pile is ready when
    a. you have added lots of garden waste.
    b. you have added lots of kitchen scraps.
    c. your pile has turned into dark, fluffy soil.

4. Compost is good for plants and flowers because
    a. it helps them grow.
    b. it keeps them warm.
    c. it keeps them dry.

5. A compost pile
    a. takes lots of room in your garden.
    b. takes several months to be ready to use.
    c. takes several years to be ready to use.

Charged-up Cars

Is there an electric car in your future?

Almost all cars, buses, and trucks run on gasoline. Gasoline is made from petroleum that has to be mined from oil wells. Gasoline isn't a very clean fuel. When cars use it, they can dirty the air. No one knows how much oil there is left, either.

What if cars could run on batteries, like a toy or a flashlight? Some of them can! Electric cars really do run on batteries. Of course, these batteries are much bigger and stronger than those used in toys. They must be plugged into a socket and charged up every night. Electric cars can't go as far or as fast as regular cars. The power stored in their batteries runs out. But electric cars are good for short trips. They don't dirty the air, either, the way gas does.

Some states and cities, such as New York City, are trying out buses that run on both gas and battery power.

QUESTIONS

1. We get _____ from oil wells.
    a. gasoline
    b. batteries
    c. petroleum
2. Gasoline isn't
    a. good for cars.
    b. a very clean fuel.
    c. bad for the air.
3. Electric cars run on
    a. batteries.
    b. natural gas.
    c. gasoline and electricity.
4. Batteries for electric cars need to
    a. run until the power runs out.
    b. be backed up by gasoline power.
    c. be plugged into a socket and charged up every night.
5. Electric cars make sense because they
    a. do not cause air pollution.
    b. take only short trips.
    c. can go very fast.

Bottle Bills Reduce Waste

What is a good way to reduce litter on highways?

In the 1970s, many people in Oregon were upset. They saw too many bottles and other litter on highways, streets, and roadsides. Litter is solid waste that is carelessly thrown away. About 40% of the litter was bottles.

The people passed a law called the Beverage Container Act. It was also known as the Bottle Bill. The law asks people to *recycle* the bottles. Reusing the bottles rather than throwing them away is known as recycling.

The Bottle Bill requires people to pay a deposit on the bottle. They get their deposit back when they return the bottle.

Oregon's Bottle Bill law was the first of its kind in the United States. The Bottle Bill was very popular. Within 2 years, litter from bottles was down by more than 80%. Within 15 years, bottles made up only 4 percent of roadside litter. Today, many states have bottle bills.

QUESTIONS

1. The word *recycle* means _____.
2. The *main idea* of the second paragraph is that
    a. litter is a major problem in Oregon.
    b. recycled bottles will help reduce litter.
    c. most litter contains bottles.
3. According to the story, today there are
    a. fewer bottles found on Oregon roads.
    b. more bottles found on Oregon roads.
    c. no bottles found on Oregon roads.
4. Which of the following statements is *true*?
    a. All states have bottle bills.
    b. Only Oregon and Maine have bottle bills.
    c. Several states have bottle bills.
5. Why do you think people need to pay a deposit when buying a bottle?
    a. People will return the bottle.
    b. People need to pay for the Bottle Bill.
    c. People will keep the bottle.

Mountain Gorillas and Dian Fossey

Mountain gorillas need to be saved.

Mountain gorillas live in high mountains of eastern and western Africa. They are the largest of the great apes and are *endangered*. *Poaching* is a major threat to the animals. Poaching is hunting in a no-hunting area. Other threats are logging and road building.

Dian Fossey, an American, set up a center in Africa to learn more about gorillas. She wanted to save them from *extinction*.

Mountain gorillas are social. They live in close family groups of 5 to 40. The adult males weigh over 300 pounds.

The males are larger than the females. They stand up about 4 feet to almost 6 feet tall.

They eat berries, roots, shoots, fruit, leaves, bark, bamboo, wild celery, and even ants. An adult male eats about 60 pounds of food each day.

Today, there are fewer than a thousand mountain gorillas in the world. However, their number has grown since the 1970s. The work done by Dian Fossey and others is helping to keep these animals from extinction.

QUESTIONS

1. Mountain gorillas live in
    a. South America.
    b. North America.
    c. Africa.
2. Mountain gorillas live
    a. alone.
    b. in pairs.
    c. in groups.
3. The *main idea* of this story is that
    a. mountain gorillas live in Africa.
    b. Dian Fossey studied mountain gorillas.
    c. mountain gorillas are in danger.
4. Mountain gorillas eat mostly
    a. ants.
    b. insects.
    c. plants.
5. In the last paragraph, what does the word *extinction* mean?
    a. no longer living on Earth
    b. living in only a few places
    c. living in many places

ENVIRONMENTAL SCIENCE

BIBLIOGRAPHY

Books on Life Science

___. *Human Body*. New York: Time-Life, 1992.

Cook, David. *Ocean Life*. New York, Crown, 1985.

Boulton, Carolyn. *Birds*, Action Science Series. New York: F. Watts, 1984.

___. *Physical Forces*. New York: Time-Life, 1992.

Elting, Mary. *Snakes & Other Reptiles*. New York: Messner, 1987.

Johnson, Sylvia A. *Crabs*. Minneapolis, MN: Lerner Publications, 1982.

Henderson, Douglass. *Dinosaur Tree*. New York: Simon & Schuster, 1994.

Lorimer, Lawrence T.; Fowler, Keith. *The Human Body: A Fascinating See-Through View of How Bodies Work*. Pleasantville, NY: Readers Digest, 1999.

Maestro, Betsy, *How Do Apples Grow?*, New York: Harper Collins, 1992.

Pfeffer, Wendy. *What's it Like to be a Fish?*. New York: Harper Collins, 1996.

Schlein, Miriam. *Project Panda Watch*. New York: Macmillan, 1984.

Showers, Paul. *What Happens to a Hamburger*. rev. ed. New York: Harper, 1985.

Silverstein, Alvin; Silverstein, Virginia; Silverstein, Laura. *Common Colds (My Health)*. New York: F. Watts, 1999.

___. *Space Planets*. New York: Time-Life, 1992.

Waldbauer, Gilbert. *Millions of Monarchs, Bunches of Beetles: How Bugs Fins Strength in Numbers*. Cambridge, MA.: Harvard Univ. Press, 2000.

Zoehfeld, Kathleen Weidner. *What's Alive?*. New York: Harper Collins, 1995.

Web-sites on Life Science

U.S. Department of Agriculture/Food and Nutrition Service
http://www.fns.usda.gov/fns

International Food Information Council
http://www.ifcinfo.health.org

Books on Earth-Space Science

___. *Rock Collecting*. New York: Crowell, 1984.

___. *Space Planets*. New York: Time-Life, 1992.

Bramwell, Martyn. *Glaciers and Ice Caps*. New York: F. Watts, 1994.

Bramwell, Martyn. *Planet Earth*. New York: F. Watts, 1987.

Bramwell, Martyn. *Rocks and Fossils*. Tulsa, OK: E D C Publishing, 1994.

Bramwell, Martyn. *The Oceans*. New York: F. Watts, 1994.

Bramwell, Martyn. *Volcanoes and Earthquakes*. New York: F. Watts, 1994.

Branley, Franklin Mansfield. *The Beginning of the Earth*. New York: Harper Row, 1988.

Cottonwood, Joe. *Quake!*. New York: Crowell, 1987.

Gans, Roma. *Danger-Iceburgs*. New York: Crowell, 1987.

Lauber, Patricia. *You're Aboard Spaceship Earth*. New York: Harper Collins, 1996.

Morris, Neil. *Earthquakes* (Wonders of Our World, No. 1). New York: Crabtree Publishing, 1998.

Pollard, Michael. *Air, Water, Weather*. New York: Facts on File, 1987.

Ricciuti, Edward R.; *Carruthers*, Margaret W. *Rocks and Minerals* (National Audubon Society). New York: Scholastic, Inc., 1998.

Tangborn, Wendell V. *Glaciers*. rev. ed. New York: Crowell, 1988.

Whyman, Kathryn. *Heat and Energy*. New York: Glouchester Press, 1986.

Web-sites on Earth-Space Science

National Severe Storms Laboratory/NOAA
http://www.nssl.noaa.gov

Books on Physical Science

___. *Physical Forces*. New York: Time-Life, 1992.

Barret, Norman. *Night Sky*. New York: F. Watts, 1986.

Barret, Norman. *Sports Machines*. New York: F. Watts, 1994.

Bramwell, Martyn. *Glaciers and Ice Caps*. New York: F. Watts, 1994.

Branley, Franklyn M. *Gravity Is a Mystery*. rev. ed. New York: Crowell, 1986.

Branley, Franklyn Mansfield. *The Beginning of the Earth*. New York: Harper Row, 1988.

Durant, Penny Raife. *Make a Splash*. New York: F. Watts, 1991.

Haslam, Andrew; Glover, David. *Machines (Make It Work! Series)*. Pittsfield, MA: World Book, Inc., 1997.

Morgan, Nina. *Lasers* (20th Century Inventions). Texas: Raintree Steck-Vaughn, 1997.

Whyman, Kathryn. *Heat and Energy*. New York: Gloucester Press, 1986.

Wyler, Rose. *Science Fun with Mud and Dirt*. New York: Messner, 1987.

BIBLIOGRAPHY

Web-sites on Physical Science

Railway Technical Research Institute
http://www.rtri.or.jp

Books on Environmental Science

Accorsi, William. *Rachel Carson*, New York, NY: Holiday House, 1993.

Berger, Melvin. *Oil Spill!*, New York: Harper Collins, 1994.

Forman, Michael H. *Artic Tundra* (Habitats). Danbury, CT: Children's Press, 1997.

George, Jean Craighead. *1 Day in the Tropical Rain Forest* (Newbery Medal Winner Series, No. 5). New York: Crowell Co., 1990.

Mongillo, John; Zierdt-Warshaw, Linda. *The Encyclopedia of Environmental Science*. Phoenix, AZ: Oryx Press, 2001.

Morgan, Sally. *Acid Rain* (EarthWatch). New York: F. Watts, 1999.

Pollock, Steve. *The Atlas of Endangered Animals* (Environmental Atlas Series). New York: Checkmark Books, 1993.

RECORD KEEPING

The Progress Charts on these pages are for use with questions that follow the stories in the Life Science, Earth-Space Science, Physical Science, and Environmental Science Units. Keeping a record of your progress will help you see how well you are doing and where you need to improve. Use the charts in the following way:

After you have checked your answers, look at the first column, headed "Questions Page." Read down the column until you find the row with the page number of the questions you have completed. Put an X through the number of each question in the row that you have answered correctly. Add the number of correct answers, and write your total score in the last column in that row.

After you have done the questions for several stories, check to see which questions you answered correctly. Which ones were incorrect? Is there a pattern? For example, you may find that you have answered most of the literal comprehension questions correctly but that you are having difficulty answering the applied comprehension questions. If so, then it is an area in which you need help.

When you have completed all of the stories in an unit, write the total number of correct answers at the bottom of each column.

PROGRESS CHART FOR LIFE SCIENCE UNIT

Questions Page	Comprehension Question Numbers				Total Number Correct per Story
	Science Vocabulary	Literal	Interpretive	Applied	
7	1	2	3,4,5		
9	1	2	3,4,5		
11	1	2,3,4	5		
13		1,2,3	4,5		
15	1	2,3	4,5		
17	1	2,3	4,5		
19	1	2,3	4,5		
21	1	2,3	4	5	
23	1	2		3,4,5	
25	1	2,3	4,5		
27	1	2,3	4,5		
29	1	2,3,4	5		
31	1	2,3	4,5		
33	1	2	3,4	5	

Total Correct by Question Type

PROGRESS CHART FOR EARTH-SPACE SCIENCE UNIT

Questions Page	Comprehension Question Numbers				Total Number Correct per Story
	Science Vocabulary	Literal	Interpretive	Applied	
39	1	2,3	4,5		
41	1	2	3,4,5		
43	1	2,3	4,5		
45	1	2,3	4,5		
47	1	2,3	4,5		
49	1	2,3	4,5		
51	1	2	3,4,5		
53	1	2,3	4,5		
55	1	2,3	4,5		

Total Correct by Question Type

RECORD KEEPING

PROGRESS CHART FOR PHYSICAL SCIENCE UNIT

Questions Page	Comprehension Question Numbers				Total Number Correct per Story
	Science Vocabulary	Literal	Interpretive	Applied	
61	1	2,3	4,5		
63	1	2,3	4,5		
65	1	2	3,4		
67	1	1,2,3	4,5		
69	1	2,3,4	5		
71	1	1,2,3	4,5		
73	1	2,3	4,5		
75	1	2,3	4,5		

Total Correct by Question Type

PROGRESS CHART FOR ENVIRONMENTAL SCIENCE UNIT

Questions Page	Comprehension Question Numbers				Total Number Correct per Story
	Science Vocabulary	Literal	Interpretive	Applied	
81	1	2,3	4,5		
83	1	2,3,4	5		
85	1	2,3	4,5		
87	1	2,4,5	3		
89	1	2,3	4,5		
91	1	2,3,4,5			
93		1,2,3,4	5		
95	1	3,4	2,5		
97	5	1,2,4	3		

Total Correct by Question Type

METRIC TABLES

This table tells you how to change customary units of measure to metric units of measure. The answers you get will not be exact.

LENGTH

Symbol	When You Know	Multiply by	To Find	Symbol
in	inches	2.5	centimeters	cm
ft	feet	30	centimeters	cm
yd	yards	0.9	meters	m
mi	miles	1.6	kilometers	km

AREA

Symbol	When You Know	Multiply by	To Find	Symbol
in^2	square inches	6.5	square centimeters	cm^2
ft^2	square feet	0.09	square centimeters	cm^2
yd^2	square yards	0.8	square meters	m^2
mi^2	square miles	2.6	square kilometers	km^2
	acres	0.4	hectares	ha

MASS (WEIGHT)

Symbol	When You Know	Multiply by	To Find	Symbol
oz	ounces	28	grams	g
lb	pounds	0.45	kilograms	kg
	short tons (200 lb)	0.9	tonnes	t

VOLUME

Symbol	When You Know	Multiply by	To Find	Symbol
tsp	teaspoons	5	milliliters	mL
Tbsp	tablespoons	15	milliliters	mL
fl oz	fluid ounces	30	milliliters	mL
c	cups	0.24	liters	L
pt	pints	0.47	liters	L
qt	quarts	0.95	liters	L
gal	gallons	3.8	liters	L
ft^3	cubic feet	0.03	cubic meters	m^3
yd^3	cubic yards	0.76	cubic meters	m^3

TEMPERATURE (exact)

Symbol	When You Know	Multiply by	To Find	Symbol
°F	Fahrenheit temperature	5/9 (after subtracting 32)	Celsius temperature	°C

METRIC TABLES

This table tells you how to change metric units of measure to customary units of measure. The answers you get will not be exact.

LENGTH

Symbol	When You Know	Multiply by	To Find	Symbol
mm	millimeters	0.04	inches	in
cm	centimeters	0.4	inches	in
m	meters	3.3	feet	ft
m	meters	1.1	yards	yd
km	kilometers	0.6	miles	mi

AREA

Symbol	When You Know	Multiply by	To Find	Symbol
cm^2	square centimeters	0.16	square inches	in^2
m^2	square meters	1.2	square yards	yd^2
km^2	square kilometers	0.4	square miles	mi^2
ha	hectares (10,000 m^2)	2.5	acres	

MASS (WEIGHT)

Symbol	When You Know	Multiply by	To Find	Symbol
g	grams	0.035	ounces	oz
kg	kilograms	2.2	pounds	lb
t	tonnes (1000 kg)	1.1	short tons	

VOLUME

Symbol	When You Know	Multiply by	To Find	Symbol
mL	milliliters	0.03	fluid ounces	fl oz
L	liters	2.1	pints	pt
L	liters	1.06	quarts	qt
L	liters	0.26	gallons	gal
m^3	cubic meters	35	cubic feet	ft^3
m^3	cubic meters	1.3	cubic yards	yd^3

TEMPERATURE (exact)

Symbol	When You Know	Multiply by	To Find	Symbol
°C	Celsius temperature	9/5 (then add 32)	Fahrenheit temperature	°F